101

AMAZING

BRAIN

TEASERS

The Ultimate Collection of Math and Logic Puzzles for Smart Kids and Teenagers

Copyright © 2021 by Unicorn Books

All rights reserved. No part of this publication may be reproduced, stored or transmitted in any form or by any means, electronic, mechanical, photocopying, recording, scanning, or otherwise without written permission from the publisher.
It is illegal to copy this book, post it to a website, or distribute it by any other means without permission.

First Edition

Table of Contents

Introduction......................................1

Brain Teasers - Part 1 3

Brain Teasers - Part 2 29

All the Answers.........................57

Your Free Gift..........................75

Introduction

This book consists of 101 challenging brain teasers. By solving them, you will strengthen your logic skills and develop your mathematical intuition. But this is not simply a list of problems. Each puzzle tells a short interesting story inspired by real-life situations and is accompanied by meaningful illustrations.

The puzzles are not organized by difficulty, and it is not necessary to follow a strict order. Flip through the pages and let yourself be inspired by the titles and images in choosing your next challenge.

This book was designed to be at the same time challenging, entertaining, and fun. The riddles don't require any advanced knowledge and use only elementary math. But don't underestimate them. They can be hard for people of any age and level of instruction. Don't believe me? Try them on your parents, your older siblings, or even your teachers.

Discussing a puzzle with others is certainly the most efficient and enjoyable way to get to the solution. For this reason, the book also represents an alternative way to spend time with friends and family, engaging them in a fun and entertaining activity.

You will notice that the more puzzles you tackle, the better you will get at solving them. That's because you are improving your ability to think outside the box, which is useful in every aspect of your life.

Before we start ...

If you wish to support our work and would like to see more books like this one available, please

leave your honest review.

And now... Let the fun begin!

Brain Teasers
Part 1

1. SECRET CODE

The code to open my padlock consists of three digits.

The sum of all three is equal to 10.
The product of the first two equals six.
The second digit is the highest.

What is the code that opens the padlock?

2. MYSTERIOUS WEIGHT

The scale below is perfectly balanced. We have the following information:
- Four of the weights measure, respectively, 2 lb, 3 lb, 5 lb, and 10 lb.
- Two out of the five weights are equal.

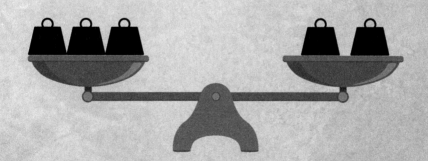

How heavy is the fifth weight?

3. PERFECT BBQ

Frank is a BBQ master. To obtain a perfect steak, he grills each side for five minutes.

Tonight, Frank wants to prepare three steaks, but unfortunately on his grill there is space for at most two steaks.

How long will it take Frank, at minimum, to grill three steaks on both sides?

4. GIGANTIC PRODUCTS

The math teacher wrote on the blackboard the following scary calculation.

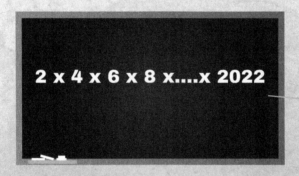

It is the product of all the even numbers from 2 to 2022. She then asks the following questions to her students.

What is the last digit of the result? And the one before last?

5. CHALLENGE FROM THE DESERT

In this piece of papyrus found in the desert, a sum was calculated.
We can distinguish the digit 1 in the result of the operation.

Find out what digits the two symbols are replacing.

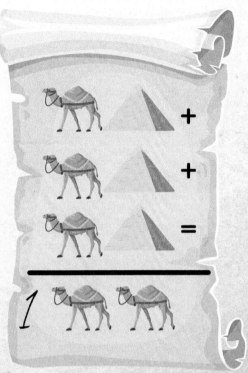

6. SPIDER'S WEBS

To move between two branches of a tree, Steve the spider built an intricate web connecting different leaves.

After a storm, Steve can still move between the two branches. **How many pieces of the web were destroyed by the storm, at most?**

7. TRIANGULAR CAT

James, who likes both geometry and animals, has produced the following drawing of a cat.

How many different triangles can you count in James's drawing?

8. CROWDED BUS

Today, when I got on the bus, I counted 13 other passengers. At the first stop, 10 more passengers got on the bus.

At the following stop, half of the passengers got off. At the stop after that, I got off, and so did 11 other passengers.

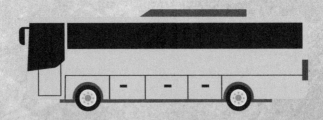

How many people were left on the bus, at that point?

9. CURIOUS BIRTHDAYS

In a group of friends, everyone has a different birthday. Still, they have something in common: to write the day and month of birth of every friend, one never needs the digits 4,5,6,7,8, or 9.

What is the maximum number of friends in that group?

10. LACK OF CASH

Marina and Robert would like to buy a new comic book. They search their pockets and find some one-dollar bills.

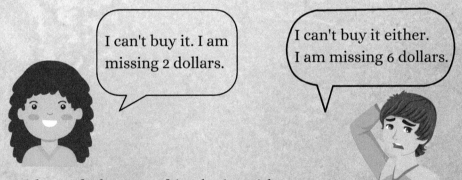

I can't buy it. I am missing 2 dollars.

I can't buy it either. I am missing 6 dollars.

Neither of the two friends is without money. However, even when they put their dollar bills together, they don't have enough money to make the purchase.

What is the price of the comic book?

11. GLOVES OBSESSION

Samantha has 60 pairs of gloves in her drawer. They are in 5 different colors, 12 pairs for each color. Since the lights are out, she is in complete darkness.

She wants to make sure she picks at least one pair of the same color.

How many gloves does she need to take out of the drawer?

12. MATHEMATICAL WONDERS

Julia has just discovered two three-digit numbers with a surprising property: they are equal to the sum of the third powers of their digits. Here is what she wrote on the blackboard to explain her discovery.

$$153 = 1^3 + 5^3 + 3^3$$
$$370 = 3^3 + 7^3 + 0^3$$

Starting from Julie's numbers, it is possible to find, without any calculations, another three-digit number with the same property.
What is this number?

13. PRESIDENTIAL RIDDLE

Simon is quoting from his trivia book. "Did you know that of the first five US presidents (in order, Washington, Adams, Jefferson, Madison, and Monroe) three died on the fourth of July?"

Lucy replies "I have never heard this before, but I am ready to bet my lunch money that Monroe is one of those three."

How can Lucy be so sure?

14. UNRELIABLE ANTS

In a remote corner of the globe, there are three types of ants. The red ones always lie but are honest on the last two days of every month. The green ants always tell the truth, except during the days ending with a zero (that is, 10, 20, or 30). In those days, they always lie. The yellow ants lie on even days and are truthful on odd days.

One day in November, I couldn't remember the exact date, so I asked three of them, "Is today the 30th?". Here are their replies.

On which day did this conversation take place?

15. PATIENCE NEEDED

I am using an old ice cream machine to fill my cone. The machine drops vanilla ice cream at a constant rate.

After five minutes, my cone is filled for half of its height.

How many more minutes will I have to wait before the cone is completely full of ice cream?

16. SELF-REFERENTIAL SENTENCE

In a monastery, the following piece of parchment was discovered. It contains an interesting self-describing sentence. Unfortunately, an ink stain is covering the first number.

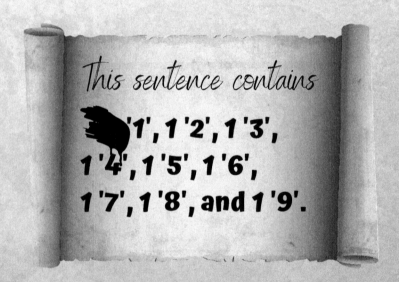

What is the number covered by the stain?

17. HUNGRY MONKEYS

Seven monkeys eat seven bananas in seven minutes.

How many monkeys will eat 98 bananas in 49 minutes?

18. STRANGE ISLANDS – PART 1

In the remote island of Sbokolk, the population is divided into two tribes: the Sboks and the Kolks. They have the strange habit of replying to any question with three sentences. But there is an important difference between the two tribes.

A Sbok always tells more true sentences than false ones, while a Kolk says more false statements than true ones.

As soon as I got to the island, I met two men, and asked them the way to the city. Here is what one of the men replied.

> I am a Kolk.
> My friend and I are both Kolks.
> You should head north to reach the city.

Should I really go north to reach the city?

19. STRANGE ISLANDS – PART 2

Before my trip to Sbokolk, I studied a bit of their language. Unfortunately, once I got there, I couldn't remember if the word for bread was "krok" or "skok". I therefore asked two local women. Here is how one of them replied.

> Between me and my friend, at least one is a Kolk.
> Only one of the three sentences I am telling you is true.
> Our word for bread is "skok".

How do you say "bread" in the island of Sbokolk?

20. STAYING HYDRATED

Daniel's bottle now weighs 1.2 lb, but it is only half full of water.

When the bottle is completely filled with water, it weighs 2 lb.

What is the weight of the empty bottle?

21. CORDIAL GATHERING

At the end of a party every participant shook hands with everyone else. There were in total 36 handshakes.

How many people were at that party?

22. SOCCER CUP

Chelsea has just won a tough game against Juventus. The final score was 7 - 5. At half-time, Chelsea had scored as many goals as Juventus later scored during the second half.

How many of the 12 total goals were scored during the first half of the game?

23. SAFE INVESTMENTS

Yesterday, the value of Albert's stocks dropped by 10%.

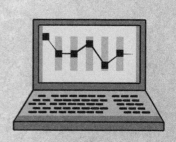

This morning he saw that it increased 10% since yesterday. He gave a sigh of relief. "Luckily, I didn't lose anything".

Is Albert correct in his assessment?

24. SONGWRITERS

Pedro and Camila form an enthusiastic music duo and only perform their own original songs.

Their repertoire consists of a total of eight songs. Pedro contributed to five of them, while Camila to six.

How many of their eight songs did they write together?

25. EVEN AND ODD

Complete the two sentences to make them true statements. (Careful because the numbers you write are also inside the rectangle).

| 1 | 2 | 3 | 4 |

Inside this rectangle there are .... odd numbers.

| 5 | | | 6 |

Inside this rectangle there are .... even numbers.

| 7 | 8 | 9 | 10 |

What numbers should be written in place of the dots?

26. AMERICAN COINS - PART 1

The coins in circulation in the United States nowadays are the penny (1¢), the nickel (5¢), the dime (10¢), and the quarter (25¢).

What is the smallest sum that cannot be paid using at most six coins?

27. AMERICAN COINS - PART 2

Martin is in front of a vending machine. With the few coins in his pockets, he wants to buy a can of soda which costs 75¢.

Despite having enough to buy the can, Martin cannot pay 75¢ exactly.

How much money does Martin have, at most, in his pockets?

28. FORBIDDEN PRICES

In a distant planet, the local currency is the Stok. There are no coins, and only three types of banknotes are printed. Their values are, respectively, 6, 9, and 20 Stoks.

Some prices, like 7 Stoks, are forbidden, since there is no way of paying them.

What is the highest forbidden price?

29. DIVERSE FRIENDSHIPS

In a group of six friends, several different languages are spoken. Given any pair of friends, each of them speaks a language that the other doesn't understand.

What is the minimum number of languages spoken in that group?

30. SWEET BREAKFAST

Yesterday, I went to a French patisserie with my friends Alyssa and Jasmine. Alyssa ordered one croissant and two cappuccinos, spending $6.60.

Jasmine got two croissants and one cappuccino. She spent $6. I ordered one croissant and one cappuccino.

How much did I spend?

31. WALL PAINTING

Levin is an expert painter. He paints an entire wall in three hours. Alex is still an apprentice, and it takes him six hours to do the same job.

Today, they started to paint a wall together.

How long will it take them to complete the task?

32. ERASE AND WRITE

On the blackboard, Lee wrote all the numbers from 1 to 10.

$$1 - 2 - 3 - 4 - 5$$
$$6 - 7 - 8 - 9 - 10$$

He now starts the following process: he chooses two numbers and erases them from the blackboard, replacing them with their sum increased by one. For instance, if he chooses 3 and 8, he erases them and writes 12 somewhere on the blackboard.

Lee keeps repeating this procedure until there is only one number left on the blackboard.

What is this number?

33. READY FOR THANKSGIVING

In a community of 20 turkey farmers, each farmer has 1, 5, or 9 turkeys.

The number of farmers who have 1 turkey is equal to the number of farmers who have 9 turkeys.

How many turkeys are raised in that community?

34. THREATENING QUEENS

You want to place three chess queens on a 3×3 chessboard so that no two queens threaten each other. That is, a solution requires that no two queens share the same row, column, or diagonal.

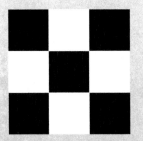

Can you solve this problem?
What if the queens are 4 on a 4x4 chessboard?

35. BALLS IN BOXES

Jesse has an impressive collection of small bouncy rubber ball., She keeps them in a cubic box whose edge measures 10 inches.

Since the box is completely full, she decides to transfer the entire collection into a bigger cubic box whose edge measures 20 inches.

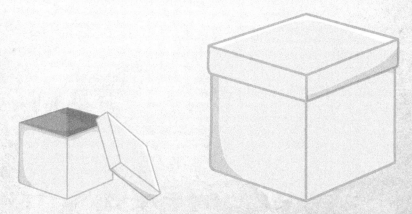

What is the height reached by the rubber balls in the second box?

36. THREE SPECIAL NUMBERS

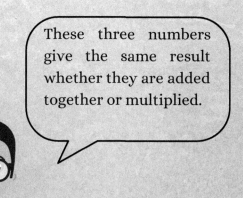

These three numbers give the same result whether they are added together or multiplied.

What are these three special numbers?

37. LOST BANKNOTE

At the library, I opened a book and a twenty-dollar bill fell out. Immediately, a woman behind me said, "It's mine. I was using it as a bookmark. I left it between the pages 124 and 125."

Another man entered the conversation, saying, "It's mine! I left it between pages 201 and 202".

I checked the book and saw that it had 246 pages.

To whom did I give the banknote?

38. THIRSTY CAMELS

Five Bactrian camels drink 5 gallons of water in 5 days.
Seven Arabian camels drink 7 gallons of water in 7 days.

Which of the two species requires a higher daily amount of water?

39. LAND DIVISION

A farmer divided his land into four equal rectangular pieces.

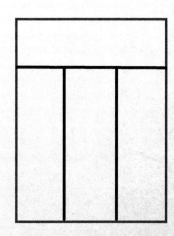

Each piece is enclosed by 160 ft of fence.

What is the total area of the land?

40. THREE CURIOUS FRIENDS

Jennifer, Linda, and Thomas are three adventurous friends looking for interesting insects and beetles in the forest.

Jennifer and Linda together found 27 insects, Thomas and Jennifer 30, and Thomas and Linda 33.

How many did they find in total? Who found more?

41. EGG EXCHANGES

Last Summer, I went to a camping adventure. There were three large groups of friends, and it was a lot of fun.

One night, Leonard, a coordinator for the third group, came to see me and Luke, who were responsible for groups number one and two, respectively.

Group number three had forgotten to bring eggs. My group had 7 dozen, and Luke's group had 4. We put all these eggs together and divided them equally among the three groups. Leonard offered us 11 dollars in exchange for the eggs.

How did we split the 11 dollars between group one and two to make the division fair?

42. MARCHING DUCKS

This morning, while he was chilling down at the lake, Tom saw two ducks marching in front of a duck, two ducks marching behind a duck, and one duck marching between two ducks.

What is the minimum number of ducks that Tom saw?

43. SPECIAL DATES

Matthew lives in England, where dates are written in the order day – month – year. Matthew notices that the date in which he was born is a palindrome, that is, it reads the same backward as forward.

20 02 2002

By a strange coincidence, Matthew's younger brother Leopold was also born in a palindrome date. Matthew notices that no palindrome dates occurred between his birth and that of his brother.

When was Leopold born?

44. THREE DOORS

In a TV show, a contestant must choose one out of three doors. Behind one of them there is the prize: a brand-new sports car. On each door there is a sign with a clue. The host of the show tells the contestant that at least one of the three clues is true, and at least one is false.

Which door should the contestant open to win the sports car?

45. SMART WEIGHING - PART 1

I have seven coins. They all look the same, but I know that one of them is fake and weighs a little less than the others.

How many times do I need to use my balance scale to find out which coin is the fake one?

46. SMART WEIGHING - PART 2

I am again faced with the problem of discovering a fake coin. This time, I have fifteen coins in total. They all look the same, but the fake one is a bit lighter than the others.

How many times do I need to use my balance scale to find out which coin is the fake one?

47. MYSTERIOUS AGES

Eleanor asks her friend Logan what the ages of his three sons are. Logan replies, "The product of the three ages is 36. Their sum is the number of the bus that is passing in front of us right now."

Eleanor stares at the bus and says, "I still don't have enough information." Logan says "My oldest has blue eyes." At that point, Eleanor knows the ages of Logan's three kids.

What is the number of the bus that is passing in front of the two friends?

48. PLENTY OF FISH

In a fishing competition, the first three caught a total of 40 fish. The second caught 50% more fish than the third. Moreover, the winner caught 4 more than double the number of fish caught by the third.

How many fish did the second catch?

49. STRANGE DICE

Peter built the following two peculiar dice.

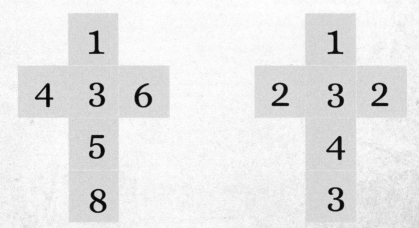

Is it easier to obtain a total of 7 by casting Peter's dice or two regular dice?

50. SPORT WITH FRIENDS

Christian would like to celebrate his promotion by going out with his friends. He is considering two different options.

He could take 10 friends to a basketball game, or he could take his 4 closest friends to 5 different hockey games. Obviously, Christian wants to pay for everyone's tickets, including his own.

Knowing that a ticket for a basketball game costs double the ticket for a hockey game, which option is cheaper?

Brain Teasers Part 2

51. CRYPTIC SEQUENCE

Sofia challenges her cousin Anthony showing him the following logic series.

MIXIE

Can you help Anthony figure out what the next symbol of the sequence should be?

52. COLD DAYS

Every day at noon, Madison checks the temperature (in degrees Fahrenheit) on her new outdoor thermometer. She then writes that number down on her notebook.

Adding together the two numbers she wrote for Monday and Tuesday, she obtains -6. Adding the numbers from Tuesday and Wednesday, she gets 48. Finally, summing together the temperatures of Wednesday and Monday, the result is 36.

After checking the thermometer on Thursday, Madison notices that the average temperature from Monday to Thursday is the same as the average temperature from Monday to Wednesday.

What temperature did Madison write down for Thursday?

53. ON AND OFF

Six light bulbs are positioned next to each other as in the picture below. Two of them are on and the others are off.

Whenever you touch one of the light bulbs, it changes its status from on to off and vice versa. However, the light bulbs immediately to the left and to the right of the one you touch also change their status.

More precisely, if you touch the first light bulb or the last one, then a total of two light bulbs will change their status. If you touch one of the four central light bulbs, then three of them will change their status at the same time.

How many light bulbs do you need to touch, at minimum, before they are all turned off?

54. JOINING TABLES

Aurora has two identical rectangular tables. Since her friends are coming over, she wants to join the tables together. If she joins them by their longer sides, she obtains a square whose perimeter is 320 inches.

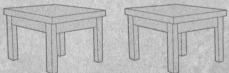

What is the perimeter of the table obtained by joining the two tables by their shorter sides?

55. DIFFICULT MEASUREMENTS

Michael wants to measure the height of his table. Unfortunately, his two favorite pets, Flippy the turtle and Bandit the cat, keep getting in the way. Here are the measurements he was able to carry out.

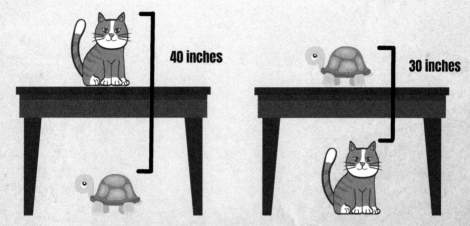

What is the height of the table?

56. ANCIENT DOCUMENT

Austin is going through an old book where an important document is hidden. He finds the following clue:

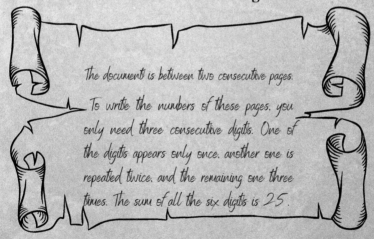

The document is between two consecutive pages. To write the numbers of these pages, you only need three consecutive digits. One of the digits appears only once, another one is repeated twice, and the remaining one three times. The sum of all the six digits is 25.

Knowing that the book has less than 400 pages, between which pages will Austin find the document?

57. SQUARE PUZZLE

Emilia has just received as Christmas present a square puzzle with 81 pieces.

She notices that some of the pieces have straight sides, while some others don't.

What is the minimum number of pieces in Emilia's puzzle that have at least one straight side?

58. MEASURING TIME – PART 1

You have two hourglasses in front of you. The first hourglass measures 7 minutes, while the other one measures 4 minutes.

How is it possible for you to measure exactly nine minutes only using these two hourglasses?

59. MEASURING TIME - PART 2

You have two ropes that are different in length and thickness. The only thing you know is that each of them burns completely in 60 minutes. You don't want to cut the ropes, since they are not homogeneous, and then you wouldn't be sure how long it would take for the different pieces to burn.

How can you measure exactly 45 minutes with only these two ropes and a lighter?

60. MATH WITH MATCHES

Alex wrote the following mathematical identity using matches. Unfortunately, he realizes that it is not correct.

Can you correct the identity by moving only one match?

61. MAKING IT EVEN

Each of the three goblets below contains some coins.

The first goblet has 6 silver coins and 5 gold ones. The second goblet has 7 gold coins. The third goblet has 6 silver coins.

One move consists in grabbing as many coins as you want (gold, silver, or a mixture of both) from one goblet, and transferring them to another goblet.

What is the minimum number of moves needed to end up with 4 gold coins and 4 silver coins in every goblet?

62. WATER TRANSFERS

You have three water containers holding respectively 3, 5, and 9 gallons of water. The biggest container is full of water, while the other two are empty.

How is it possible for you to end up with exactly 4 gallons of water in the 5-gallon tank?

63. OLD SCHOOL

Lord Stratford uses a sumptuous ancient carriage to travel the 42 miles separating his house in London from his estate in the countryside.

Unfortunately, the carriage is so old that each of the four wheels needs to be replaced after 24 miles.

The Lord is getting ready to leave from London when he notices that the new coachman, Mr. Redruth, is only bringing three additional wheels for the trip. - It is alright, My Lord - says Redruth - we will manage it.

What does the coachman have in mind?

64. YELLOW AND BLUE

In a bag there are three coins. One is yellow on both sides, the other one is blue on both sides. The third one has one blue side and one yellow side.

With your eyes closed, you pick one coin from the bag and place it on the table in front of you. After opening your eyes, you see that the upward side is blue.

If you had to bet on the color of the side touching the table, what would you say?

65. CHESS CLUB

At the first meeting of the school chess club, 5 students show up. They realize that each of them is already friend with either 2 or 3 other club members.

Moreover, when two members are friends, they don't have the same number of friends in the club. Philip is friend with Aaron and Riley. Sarah has three friends in the club.

Who are Deborah's friends in the club?

66. COOKIES AT GRANDMA'S

On Saturday, Gabe's grandma bakes cookies, always the same number. Gabe and his siblings divide the cookies equally among themselves.

Last Saturday, Gabe's brother Micky was missing. As a result, each child got two more cookies. Today, since Gabe brought a friend, each child receives one less cookie than usual.

How many siblings does Gabe have, at minimum?

67. WINE ARRANGEMENT

In front of you there are six wine glasses. Three of them are empty, while three contain wine.

You would like to arrange them so that no empty glass is next to another empty glass and no glass with wine is next to another glass with wine.

What is the minimum number of glasses you need to touch to achieve this goal?

68. COLORFUL FLAG

The flag of a newly founded country is a rectangle divided into four squares of different colors.

The total area of the flag is 60 square feet.

What is the measure of the perimeter of the flag?

69. MARKET DAY

At the local market, one can exchange one duck with two chickens. One cow can be exchanged with one goat and three ducks. Finally, one goat is worth two ducks and two chickens.

George the farmer is bringing two cows to the market.

What is the maximum number of chickens that George can be expected to bring home?

70. FOUR SWITCHES

Outside a room there are four switches, all in the on position. The switches control four different lightbulbs inside the room. Your goal is to match each lightbulb with the corresponding switch.

You can operate the switches in any way you like, but you have no way of seeing what happens inside the room. You can only enter the room once, and you won't be able to touch the switches again after you enter the room.

How will you carry out this task?

71. TIC-TAC-TOE

Ellie and Nora are playing a game of Tic-Tac-Toe. Ellie starts by drawing an X on the bottom left corner. Nora then draws a circle in the square on the right edge.

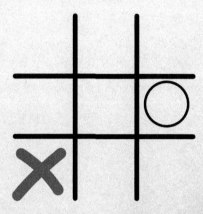

It's Ellie's turn again, and she is thinking how to secure the game.
Which moves would make Ellie 100% certain to win the game?

72. SUBTLE DIFFERENCES

Samuel has collected books for his entire life, and now he wants to sell some of them. He chooses 60 books and makes two piles of 30 books each. On the two piles, Samuel puts the following signs:

2 for $1   **3 for $1**

To his delight, during the day he sells all the 60 books.
The following day, Samuel has another batch of 60 books to sell. He thinks: "2 books for $1 and 3 books for $1 is the same as 5 books for $2". So, he creates a unique pile of 60 books with the sign:

5 for $2

Once again, he sells all the 60 books.
Did Samuel earn the same amount as the day before? Why?

73. SKETCHY HOUSE

Patrick has challenged his sister Leah to make the famous simple drawing of a "house" without lifting the pencil from the paper or passing more than once over the same line.

Leah swiftly accomplishes the task. Now she marks every corner of her drawing with a circle and asks her brother: "From how many of these five points could I have started to make the drawing following your rules?"

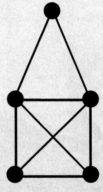

Can you help Patrick find the answer to his sister's question?

74. PIRATE LIFE

Seven pirate ships, all equal, leave together for a dangerous mission. Each ship only carries two thirds of the maximum number of pirates it can transport.

One fourth of the total number of pirates perish during the mission.

How many ships are needed, at least, to bring back all the surviving pirates?

75. IMPOSSIBLE CHALLENGE

Liam is visiting his grandfather Walter in the countryside. He challenges him in the following way. "Do you think you could make four circular enclosures with your fence so that in each of them there is at least one pig? Oh, and each of the enclosures must also contain an even number of pigs".

Walter stares at his six pigs for a moment, then brightens up. "Of course, son, nothing is impossible here. I would just need the four circular enclosures to be of different sizes".

What does Walter have in mind?

76. PREDICTING THE FUTURE

Dylan and Elena meet an old man who claims he can correctly answer every YES/NO question they will ask him.

Dylan is skeptical and is racking his brain searching for some unusual fact on which he can test the man.

Elena says "It is not necessary, we can prove him wrong just using logic".

What will Elena ask the man?

77. COUNTING CABINS

To reach the top of my favorite mountain, one must use an aerial tramway.

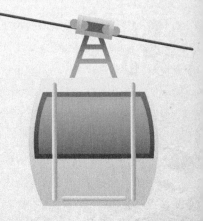

The tramway has numbered cabins, evenly spaced throughout the line. Each cabin reaches the top of the mountain and then makes its descent along the same side of the mountain.

One day, I noticed that while cabin 9 was crossing path with cabin 15 near the top of the mountain, cabin 25 and cabin 29 were meeting in the valley.

How many cabins are there in total in that aerial tramway?

78. CLOCK GEOMETRY

Today at school, Nicholas learned about right angles.

When he gets home, Nicholas stares at the big clock in the living room. He notices that, at 3 pm exactly, the minute hand and the hour hand form a right angle.

The same thing will happen again at 3 am. Nicholas imagines staring at the watch for all those 12 hours.

How many times would Nicholas see the minute hand and the hour hand forming a right angle (including the initial and final moment, that is, 3 pm and 3 am)?

79. CHRISTMAS MATH

Greta loves Christmas almost as much as she loves mathematics.

She decides to decorate the balls that she will hang on her Christmas tree by writing on them some numbers in letters. In doing so, she follows a very precise logic pattern. Here are the first five balls she decorated, in order.

At this point, Greta stops, perplexed. She is considering changing the pattern.

Can you understand why?

80. FUN TRAINING

Scarlett is training for a running competition on a circular track. With her is her dog Zipper. The two start running at the same time, but it takes Zipper only 3 minutes to run a full lap, while Scarlett needs 4 minutes.

After how many minutes will Zipper have run one full lap more than Scarlett?

81. PREMIUM SALAD

For her homemade Greek salad, Vasiliki bought four ingredients. Without the cucumbers, she would have spent $11, without the feta cheese $8, without the tomatoes $10, and without the kalamata olives $7.

How much did Vasiliki spend for her traditional Greek salad?

82. INACCURATE CLOCKS

Both clocks in Stella's rooms are inaccurate. The first one runs fast and gains five minutes every hour. The second clock does the opposite. It runs behind and loses 5 minutes every hour.

Stella's mom set the clocks to the right time earlier today. Now it is afternoon, and here is what the two clocks are showing.

At what time did Stella's mom set the two clocks?

83. OF MICE AND CHEESES

In the three situations below, the scale is always perfectly balanced.

How many mice are needed to match the weight of the sack?

84. CUBIC CALENDAR

Edgar hast two nice wooden cubes. He wants to write a digit on each face of the cubes in order to use them as a calendar. More precisely, he wants to be able to display with the cubes all the numbers from 01 to 31.

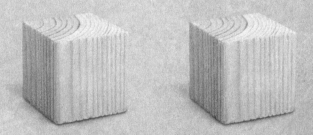

What digits should Edgar write on the faces of the two cubes?

85. TRICKY PERCENTAGES

Eddy has just cut a slice from a sweet, juicy watermelon. Unfortunately, before she has a chance of tasting it, her father calls her to attend to an urgent matter.

It is a hot summer day, and when Eddy gets back, her slice of watermelon is spoiled.

Initially, the slice of watermelon weighed 1 lb and was 99% water. Now, after an afternoon under the sun, the slice is 98% water since some of it has evaporated.

What is the weight of Eddy's slice of watermelon now?

86. BIKE EXCURSION

Eliana and Stefan went for a bike excursion. They left home at 8 am and got back at 11.30 am.

"Next time we should try to not stop on the way" says Stefan. Eliana replies, "You are right. But we did well anyways. During every 60-minute interval of our excursion we covered exactly 4 miles".

What distance did Eliana and Stefan cover, at most?

87. IMPERFECT DICE

William's two new dice are identical, but the numbers on the faces were printed incorrectly. That is, they don't satisfy the usual requirement that the sum of the numbers on opposite faces is always 7.

William casts the dice. The sum of the two numbers obtained is 7. The sum of the numbers on the two faces that are touching the table is 4.

What number is written on the face opposite to 6?

88. TRICKY COINS

Ralph was given by his grandfather the six coins below. Each of them is marked with a number.

Ralph's task is to find the coins whose sum is 50. Ralph thinks that the problem is impossible, but his grandfather keeps chuckling and insisting that a solution exists.

Which coins is Ralph's grandfather referring to?

89. PROFITABLE HELP

Every time I help my mother wash the dishes, I get 7 dollars. Every time my older brother helps my mom wash the dishes, he gets only 3 dollars. I don't find it fair, so at the end of the week we always put together all the money we have received and then each of us takes half of the total.

This week, each of us ended up 16 dollars richer from helping in the kitchen.

How many times did I wash the dishes with my mother?

90. BUSINESS ENTERPRISE

At a country fair, I bought 20 bottles of good wine for $5 each. I then started reselling them, trying to make a profit. The first day, I asked $9 per bottle, the second day $8, the third day $7 and so on, always lowering the price by $1 every day.

I sold 4 bottles the first day and 8 bottles the second day. By the end of the fourth day, I had no bottles left and I had made a total profit of $54.

How many bottles did I sell on the third day?

91. ENIGMATIC AGE

My grandfather is an honest man, and always tells the truth. Sometimes, however, he is not very straightforward. For instance, yesterday someone asked him his age. He replied in the following way.

> If my age is divisible by two, then it is between 50 and 59.
> If my age is not divisible by 3, then it is between 60 and 69.
> If my age is not divisible by 4, then it is between 70 and 79.

What is my grandfather's age?

92. PECULIAR PETS

Russell likes to surround himself with strange animals. In his living room we can find crabs, snakes, and spiders. In total, Russell can count 7 heads and 28 legs (remember that a spider has 8 legs, while a crab has 10).

How many animals does Russell have of each species?

93. TIMELY ARRIVAL

It takes Max 24 minutes to walk from his house to the train station. If he runs, instead, it takes him only 12 minutes to cover the same distance.

One morning, he leaves his house at 8 am, with plenty of time to walk to the station and catch his 8.30 am train. On his way to the station, he realizes he forgot his wallet at home. He thus runs back home and then all the way to the train station. He gets there exactly at 8.30 am, just in time to catch the train to work.

At what time did Max realize his wallet was missing?

94. CRACK THE CODE

The PIN code to unlock my phone consists of four digits, all different from each other. The first one is the product of the other three and is also strictly bigger than the sum of the other three. Moreover, the four digits are in decreasing order.

What is the PIN code of my phone?

95. FAIR TEACHER

Ms. Rush is a dedicated math teacher. For today's class, she has prepared a quiz consisting of 5 questions. Here is how she is going to assign points for each question.

> Correct Answer = +4 points
>
> No Answer = +1 point
>
> Wrong Answer = 0 point

Ms. Rush is 100% sure that in her class there will be at least two students with a different score.

What is the minimum number of students in that class?

96. FAIR ROBBERS

The bank robbery was successful, and the thieves are now in possession of a considerable number of sacks full of dollars.

One of them says: "I think Dick, who was the last to join the group, should receive one sack, then Ned will get two, Giuliano three, and so on. This way, we will distribute all the sacks".

The head of the group disagrees. "Each of us will take the same number of sacks". So they did, and each thief ended up with 5.

How many thieves are there in that group?

97. MATH AND HISTORY

The year in which James A. Garfield, lawyer and amateur mathematician, became U.S. president was a palindrome, that is, a number that reads the same backward or forward. Moreover, it was also a multiple of 3.

In which year did Garfield become president?

98. GRANDMA'S JAM

My grandma gave me three jars of handmade jam: one of strawberry jam (my favorite), one of raspberry jam, and one of apricot jam. To test my logic skills, she says: "Whenever you see two labels on a jar, only one of them tells the truth".

What type of jam is contained in the smallest jar?

99. SECRET AGENT

Misty works as a secret agent, and she is currently trying to infiltrate the headquarters of a foreign enemy.

She was able to deactivate the alarm, but she needs to act fast before her presence is discovered. Unfortunately, there is one last obstacle to overcome to open the door to the caveau containing the most important documents.

A touch-screen computer connected to a self-powered generator appears in front of Misty. It's a grid with nine squares. A question mark is flashing on the last one.

It's a secret code! And she only has a few minutes to crack it. Impatient, she looks at her digital clock, and has a sudden flash of inspiration.

What symbol will Misty draw in place of the question mark?

100. NOTEWORTHY COLLECTION

This year's new set of animal cards consists of 190 cards. Maya has nearly completed the collection. She is now trying to organize her cards to see how many she is missing.

When she divides her cards into piles of 4, she has 3 cards left. When she divides them into piles of 5, she has 4 left. Finally, when she divides her cards into piles of 6, she has 5 left.

How many cards does Maya need to complete her collection?

101. PENALTY SHOOTOUT

Asher, Jackson, and Mateo shoot 10 penalties each. Oliver is a strong goalkeeper and manages to save half of the shots.

Here is what Asher has to say at the end. "I scored 4 more goals than Mateo. One between Mateo and Jackson scored 7 goals".

How many goals did each of the three friends score?

All the Answers

1. SECRET CODE

The code is the following

2. MYSTERIOUS WEIGHT

The fifth weight measures 10 lb.

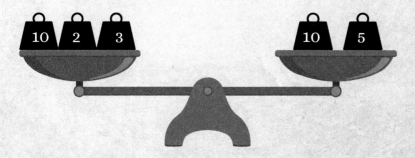

3. PERFECT BBQ

It takes Frank 15 minutes to grill the three steaks. Let us call the steaks A, B, and C. Then
5 minutes: first side of A, first side of B;
5 minutes: second side of A, first side of C;
5 minutes: second side of B; second side of C.

4. GIGANTIC PRODUCTS

Since among the factors there are the numbers 10, 100, and 1000, the product will end with quite a few zeros.

5. CHALLENGE FROM THE DESERT

6. SPIDER'S WEBS

Here is the shortest path between the branches. The storm destroyed at most 11 pieces of the web.

7. TRIANGULAR CAT

You can count a total of 17 triangles.

8. CROWDED BUS

Only one person was left on the bus: the driver.

9. CURIOUS BIRTHDAYS

There are at most 63 friends in the group.

10. LACK OF CASH

The comic book costs $7. Marina has 5 dollars and Robert one.

11. GLOVES OBSESSION

Samantha must pick at least 31 gloves. The worst-case scenario is when she picks all 30 right (or left) gloves.

12. MATHEMATICAL WONDERS

The number is 371.

13. PRESIDENTIAL RIDDLE

If Monroe hadn't died on July 4, then he wouldn't have been included. In that case, Simon would have read in his trivia book that, of the first **four** US presidents, three died on the fourth of July.

14. UNRELIABLE ANTS

The conversation took place on November 30.

15. PATIENCE NEEDED

I must wait 7 more minutes. In fact, by filling the cone to half its height, the machine has filled a smaller cone whose volume is one eighth of the total volume.

16. SELF-REFERENTIAL SENTENCE

There are two solutions. The stain could be covering the number 10 or the number 11.

17. HUNGRY MONKEYS

The answer is 14 monkeys. From the riddle we understand that one monkey eats a banana in 7 minutes. Therefore, one monkey eats 7 bananas in 49 minutes.

18. STRANGE ISLANDS – PART 1

No, going north is not a good choice.

19. STRANGE ISLANDS – PART 2

The word for "bread" is "skok".

20. STAYING HYDRATED

The empty bottle weighs 0.4 lb.

21. CORDIAL GATHERING

There were 9 people at the party.

22. SOCCER CUP

During the first half of the game 5 goals were scored.

23. SAFE INVESTMENTS

The value of Albert's stocks dropped by 1%.

24. SONGWRITERS

Pedro and Camila wrote 3 songs together.

25. EVEN AND ODD

1	2	3	4
	Inside this rectangle there are 7 odd numbers.		
5			6
	Inside this rectangle there are 5 even numbers.		
7	8	9	10

26. AMERICAN COINS – PART 1

The smallest sum that cannot be paid using at most six coins is 44 cents.

27. AMERICAN COINS - PART 2

At most, Martin has 94 cents in his pocket.

28. FORBIDDEN PRICES

The highest forbidden price is 43 Stoks.

29. DIVERSE FRIENDSHIPS

The minimum number of languages spoken by the six friends is four (each of them speaks two).

30. SWEET BREAKFAST

I spent $4.20 for my croissant and cappuccino.

31. WALL PAINTING

It will take Levin and Alex two hours to paint the wall.

32. ERASE AND WRITE

The number left on the blackboard is 64. In fact, 64 is the sum of the numbers from 1 to 10 (1+2+3+...+10=55) plus 9, which in turn is the number of times the cancellation has been performed.

33. READY FOR THANKSGIVING

Suppose that each farmer who has 9 turkeys gave 4 turkeys to a farmer who only has one. Then every farmer in the community would have 5 turkeys. In this way, we can establish that there are 5x20=100 turkeys.

34. THREATENING QUEENS

On a 3x3 chessboard the problem has no solutions. Here is a possible configuration for the 4x4 case.

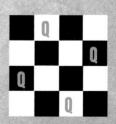

35. BALLS IN BOXES

In the bigger box, the level of the balls reaches a height of 2.5 inches.

36. THREE SPECIAL NUMBERS

The special numbers are 1, 2, and 3.

37. LOST BANKNOTE

I gave the banknote to the woman. In fact, it is impossible to stick a banknote between the pages 201 and 202 of a book.

38. THIRSTY CAMELS

A Bactrian camel drinks one gallon of water every 5 days, while an Arabian camel drinks one gallon of water every 7 days. Therefore, a Bactrian camel has a higher daily consumption of water.

39. LAND DIVISION

From the picture, one sees that in every rectangular piece the long side is three times the short side. Therefore, the perimeter is equal to 8 times the short side. Hence the short side measures 20 ft, the long side 60 ft, and the total area is 20x60x4=4800 square feet.

40. THREE CURIOUS FRIENDS

If you sum the three given numbers, the contribution of each friend is counted twice. Therefore, the total number of insects found by the three friends is (27+30+33)/2=45. Thomas is the one who found more, since his name appears in the two highest partial sums given in the problem.

41. EGG EXCHANGES

We contributed with 7x12=84 eggs. Group 2 contributed with 4x12=48 eggs. Therefore, there were in total 84+48=132 eggs. Each group received 132/3=44 eggs. Group 2 only lost 4 eggs, while we lost 40. Hence, we took 10 of the 11 dollars and left them 1 dollar.

42. MARCHING DUCKS

Tom saw at least 3 ducks.

43. SPECIAL DATES

Leopold was born on the first of February in the year 2010.

44. THREE DOORS

The contestant should open the third door to find the car.

45. SMART WEIGHING – PART 1

It is enough to use the scale twice. In the first weighing, one puts 3 coins on one plate and 3 on the other. If they balance, then the fake coin is the seventh one. If they don't, then one considers the three coins whose total weight is less and performs a second weighing, putting one coin in each plate, and leaving one out.

46. SMART WEIGHING – PART 2

There are several solutions, both requiring using the scale three times. A simple one is weighing 7 coins against other 7 to reduce the examination to a group of 7, among which the fake coin can be found. Then one proceeds as in the previous problem.

47. MYSTERIOUS AGES

Here are all the possible triples of numbers whose product is 36. Next to every triple we indicate the sum of the three numbers.

36,1,1   (38)
18,2,1   (21)
12,3,1   (16)
9,4,1   (14)
9,2,2   (13)
6,6,1   (13)
6,3,2   (11)
4,3,3   (10)

The only two triples with the same sum are 9,2,2 and 6,6,1. Therefore, the bus number is 13. Since Logan talks about "his oldest", Eleanor deduces that the ages of his children are 9,2,2.

48. PLENTY OF FISH

The second caught 12 fish. Moreover, the first caught 20 fish and the third 8.

49. STRANGE DICE

You have the same probability to obtain 7 with Peter's dice as you have with regular dice. Indeed, casting two regular dice, there are 6 ways to obtain 7: 1 - 6, 2 - 5, 3 - 4, 4 - 3, 5 - 2, 6 - 1. With Peter's dice there are also 6 ways: 3 - 4, 4 - 3, 4 - 3, 5 - 2, 5 - 2, 6 - 1. Note that some pairs are repeated because in Peter's dice some numbers appear multiple time on the same die.

50. SPORT WITH FRIENDS

It is cheaper to take 10 friends to a basketball game.

51. CRYPTIC SEQUENCE

The sequence consists of the first letter of each day of the week together with its reflection. The next symbol in the sequence is therefore an S surmounting an inverted S.

52. COLD DAYS

The temperature on Thursday was the same as the average temperature of the three previous day. To compute that average, it is enough to divide by three the sum of the three temperatures. This in turn is equal to (-6+48+36)/2=39, since by adding together the three given number the temperature of each day is counted twice. Therefore, the temperature we are looking for is 13F.

53. ON AND OFF

One must touch 4 light bulbs in the order described below.

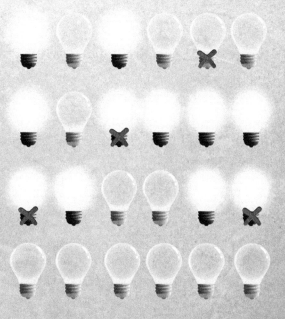

54. JOINING TABLES

The perimeter is 400 inches.

55. DIFFICULT MEASUREMENTS

The height of the table is 35 inches.

56. ANCIENT DOCUMENT

The document is between pages 354 and 355.

57. SQUARE PUZZLE

The minimum number of pieces having at least one straight side is 32.

58. MEASURING TIME – PART 1

Start the two hourglasses at the same time. After 4 minutes, turn the 4-minute one upside down. Do the same after 7 minutes with the other one. When the 4-minute hourglass stops for the second time, 8 minutes have passed. The 7-minute hourglass has been running for only 1 minute. Therefore, it is enough to turn it upside down to measure one additional minute.

59. MEASURING TIME – PART 2

Light the first rope on one end and the second rope on both ends. After 30 minutes, the second rope will be completely burned. At that point, light the other end of the first rope. It will take 15 additional minutes to burn completely.

60. MATH WITH MATCHES

Remove one match from the 9 to turn it into a 4. The identity becomes 2+2=4.

61. MAKING IT EVEN

You need three moves. For instance:
1) Transfer 4 silver coins and 1 gold coin from the first to the second goblet.
2) Transfer 4 gold coins from the second to the third goblet.
3) Transfer 2 silver coins from the third to the first goblet.

62. WATER TRANSFERS

Here is a possible solution. Fill the 3-gallon container and then empty it into the 5-gallon one. Repeat the procedure a second time. You are left with 1 gallon in the 3-gallon container. Throw away the water in the 5-gallon container and fill it with the leftover gallon plus 3 gallons.

63. OLD SCHOOL

Redruth will stop every 6 miles to rotate the wheels so that each wheel is used for only 24 miles. Let us call the 7 wheels A,B,C,D,E,F, and G. Here is a possible solution.

>   Miles 0 to 6:   A - B - C - D
>   Miles 6 to 12:   E - A - B - C
>   Miles 12 to 18:   F - E - A - B
>   Miles 18 to 24:   G - F - E - A
>   Miles 24 to 30:   D - G - F - E
>   Miles 30 to 36:   C - D - G - F
>   Miles 36 to 42:   B - C - D - G

64. YELLOW AND BLUE

You should bet on the second side of the coin being blue.

65. CHESS CLUB

Deborah's friends in the club are Sarah and Philip.

66. COOKIES AT GRANDMA'S

At minimum, Gabe has 2 siblings. The three children usually receive 4 cookies each.

67. WINE ARRANGEMENT

It is enough to touch one glass: the one before last. You pick it up and pour its content into the second empty glass.

68. COLORFUL FLAG

Note that the flag contains 15 times the smallest square. Therefore, the area of the small square is 4 square feet.
The perimeter measures 32 feet.

69. MARKET DAY

George brings home at most 24 chickens.

70. FOUR SWITCHES

Turn off two of the switches, wait for a while and then turn one of them back on. At the same time, turn off one of the two switches you had not touched before. When you enter the room, the four light bulbs can be matched to the switches according to their properties: two light bulbs will be on and two off, and in each pair, one will be hot and one cold.

71. TIC-TAC-TOE

Occupying either the bottom right corner or the top left corner would make Ellie 100% certain to win the game.

72. SUBTLE DIFFERENCES

Samuel earns one dollar less than the previous day. In fact, what Samuel does the second day is equivalent to selling 12x3=36 books at $1 every 3 and 12x2=24 books at $1 every two.

73. SKETCHY HOUSE

Leah could only have started from the two corners at the bottom.

74. PIRATE LIFE

At least four ships are needed to bring back all the surviving pirates.

75. IMPOSSIBLE CHALLENGE

Walter will just build an enclosure around the four pigs and then build the other enclosures around that first one.

76. PREDICTING THE FUTURE

Elena asks the man "Will your next answer be NO?"

77. COUNTING CABINS

There are 30 cabins in the aerial tramway.

78. CLOCK GEOMETRY

Nicholas would see the hour hand and the minute hand form a right angle 23 times.

79. CHRISTMAS MATH

The number written on every ball starting from the second one is equal to the number of letters of the number of the previous ball. Unfortunately, Greta is now stuck, because she would have to keep writing "four" in every subsequent ball.

80. FUN TRAINING

After 12 minutes.

81. PREMIUM SALAD

In the sum of the four given numbers, every ingredient is counted three times. Therefore, the total price of the salad was (11+8+10+7)/3=12 dollars.

82. INACCURATE CLOCKS

Stella's mom set the two clocks at 4 am.

83. OF MICE AND CHEESES

Three mice are needed to match the weight of the sack.

84. CUBIC CALENDAR

The trick is to use the 6 to display both 6 and 9. Here is what should be on the faces of the two cubes.
First cube: 0-1-2-3-4-5
Second cube: 0-1-2-6-7-8

85. TRICKY PERCENTAGES

The slice of watermelon now weighs 0.5 lb.

86. BIKE EXCURSION

They covered at most 16 miles (in case they went at the constant speed of 8 miles per hour during the first 30 minutes of every hour, then rested for the next 30 minutes).

87. IMPERFECT DICE

On the face opposite to 6 there is the number 4.

88. TRICKY COINS

Ralph's grandfather is referring to the three coins displayed to the right (the 6 can also be seen as a 9).

89. PROFITABLE HELP

I washed the dishes with my mother twice.

90. BUSINESS ENTERPRISE

On the third day I sold six bottles.

91. ENIGMATIC AGE

My grandfather is 75 years old.

92. PECULIAR PETS

Russell has two crabs, one spider, and four snakes.

93. TIMELY ARRIVAL

Max realized that his wallet was missing at 12.12.

94. CRACK THE CODE

The PIN code of my phone is 841.

95. FAIR TEACHER

All the numbers from 0 to 20 are possible final scores except 15, 18, and 19. Therefore, there are 18 possible scores. The minimum number of students in the class is 19.

96. FAIR ROBBERS

There are 9 thieves in the group.
In fact, 9x5=45=1+2+3+4+5+6+7+8+9.

97. MATH AND HISTORY

James A. Garfield became US president in 1881.

98. GRANDMA'S JAM

The smallest jar contains strawberry jam.

99. SECRET AGENT

In every rectangle, the segments shown are those that are normally turned off when seeing the numbers from 1 to 9 as displayed in a led digital clock. Therefore, the missing symbol consists of the single segment that is turned off when the number 9 is displayed.

100. NOTEWORTHY COLLECTION

Maya needs 11 more cards to complete her collection.

101. PENALTY SHOOTOUT

Mateo scored 2 goals, Asher 6, and Jackson 7.

Your Free Gift

We hope you enjoyed this collection of brain teasers.

As a way of saying thank you for your purchase, we would like to offer you for FREE the ebook **12 Tricky Brain Teasers to enjoy with Family and Friends.**

Scan this QR code to receive your free gift now!

Other Books in this Series

Make sure to check out on Amazon the other volumes of the series *Logic Games for Smart Kids and Teenagers*.

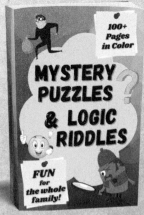

Printed in Great Britain
by Amazon